The Mathematics IA
Earning Full Marks on HL or SL Mathematics Explorations
Ideal for the International Baccalaureate Diploma

by Daniel Durwood Slosberg

2012-20 Edition

First Exams May 2014
Last Exams November 2020

©2013 by Mathematics Publishing Propriety Limited
©2015 by Rainbowdash Publishers LLC
All rights reserved.

Mr. Slosberg

National Library Catalogue-in-Publication Data

Slosberg, Daniel Durwood, 1977 - .
The Mathematics IA: Earning Full Marks on HL or SL Mathematics Explorations:
Ideal for the International Baccalaureate Diploma.

2012-20 Edition
For secondary school students.

Kindle ASIN: B00U4UE8EA
Apple ID: 1099569995; Apple Vendor ID: 13039100815
B&N Identifier: 2940158382765
Paperback ISBN-13: 978-1520901329

1. Mathematics -- Textbooks. 2. International Baccalaureate.
I. Title

2012-20 Edition first published 2013 on Kindle by Mathematics Publishing Pty Ltd
2012-20 Edition republished 2014 on CreateSpace by Mathematics Publishing Pty Ltd
2012-20 Edition republished 2014 on CopySafe by Mathematics Publishing Pty Ltd
2012-20 Edition republished 2015 on Kindle by Rainbowdash Publishers LLC
2012-20 Edition republished 2016 on iBooks and Nook by Rainbowdash Publishers LLC
2012-20 Edition paperback published 2017 on KDP by Rainbowdash Publishers LLC

Copyright ©2013 Mathematics Publishing Pty. Ltd.
Copyright ©2015 Rainbowdash Publishers LLC.

No part of this publication may be reproduced, stored in a retrieval system or transmitted in any form, or by any means, without the prior permission of the publishers.

While every care has been taken to trace and acknowledge copyright, the publishers tender their apologies for any accidental infringement or unconscious copying that has occurred or where copyright has proved untraceable. In any case where it is considered copying has occurred, the publishers would be pleased to come to a suitable arrangement with the rightful owner.

This publication has been developed by the author in his personal capacity independently of the International Baccalaureate Organization and this publication is not endorsed by that organization.

Published by Rainbowdash Publishers LLC
eMail address: RainbowdashPublishersLLC@umich.edu

IF YOU SPOT AN ERROR

We at Rainbowdash Publishers LLC feel that accuracy in our publications is of the utmost importance. This is doubly important in a mathematics textbook where an incorrectly displaying equation can cause needless confusion for the student. If you ever see an incorrect equation, kindly help us by taking a screen shot of your Kindle to show us what went wrong and eMailing it to us. Please include the Purchase ID (below), the make, model, and version of your Kindle (i.e. iPad 2 iOS 6.13 Kindle v3.9), the equation that was incorrect and what you believe should be changed to make it correct, and the kindle location number where you saw the error. We will correct the problem as soon as possible and push the correction out to all our customers.

The IB occasionally modifies its requirements between syllabus updates. If you spot something which is now out of date, please let me know so that I can update all electronic versions of this book and ensure that any new printed copies have the correction incorporated. It is helpful if you include the document and page number where you saw the discrepancy.

Thank you for your patronage and thank you in advance for letting us know about any errors that you may spot.

-Daniel D. Slosberg
RainbowdashPublishersLLC@umich.edu

Purchase ID#20160505mia

Mr. Slosberg

The Mathematics IA

2012-20 Edition

TABLE OF CONTENTS

If You Spot an Error iii
Table of Contents v
Dedications 1
Notes to Teachers on the Exploration 3
Notes to Students on the Exploration 5
Brainstorming 7
Personal Selection 9
Mathematical Selection 11
Critical Reflection 13
Teacher Conferences 15
The Abstract 17
    Jessi's Abstract   18
    Analysis of Jessi's Abstract   19
Rough Draft 23
Polishing 27
Bibliography 29
About the Author 31
Earning Full Marks Books 33
Mr. Slosberg's Other Books 34
Other Rainbowdash Authors 34

DEDICATIONS

In Memory of Peter Smythe 1943 - 2014.

To Monica, Selena, and Rebecca, who all made time for Daddy to write his first book.

To Gramps, who inspired me to become an author.

To Peter & Jenny Smythe who gave me the impetus to write this book.

To Maggie Koong, who graciously waived clause 13 in my contract to allow me to work on this book while teaching at Victoria Shanghai Academy.

To my students, and especially to Jessi who provided her expertly written abstract as a sample for future students to follow and Rahim who contributed a sample reflection.

To my fellow teachers who critiqued what I have written and made it better, especially Natasha Williams who gave many specific suggestions which I incorporated.

Mr. Slosberg

NOTES TO TEACHERS ON THE EXPLORATION

The exploration should be started with a brainstorm soon after the course begins so students can begin thinking about what they are going to research and so they will pay special attention when the parts of the syllabus they will use are taught. Students can create mind maps in groups that can be hung around the classroom to allow the students to think about them and let their ideas gel over time.

Students should select their specific topics somewhat later once they have had a chance to explore the topics in lectures and homework. In my classes, I usually do this sometime in the first term of the first year of DP. This gives students enough time to try out a few mathematical ideas to see which ones work. Some students will have enough math to begin early on. Others will need topics from future lessons. Some SL students may even want to explore HL topics or HL Math students may want to explore HL Further Math topics not covered in your HL Math course. If your class size is small or the needs of your students are clustered, you might consider teaching topics in an order tailored to them so that the students can begin working on their explorations as soon as possible. In HL Math you might even choose which option(s) to do based on what would help students' explorations the most.

By the end of term 1 in year 1, I try to make sure all students have the math they need to successfully complete their explorations (§3) -- either because it is in the course and I have taught it, or because it is outside the course but I have verified that they are successfully learning it on their own. The larger the class size, the harder it is to do this.

Conferencing with students (§4) is vital to having a class where everyone has a successful outcome. Several of the 10 hours (HL Mathematics Guide, 2012, pp. 10 & 67; SL Mathematics Guide, 2012, pp. 10 & 46) the IBO dedicates to the exploration should be spent in conferences.

It is my experience that many students struggle with the art of reflecting throughout their exploration (§5). Many students want to write the entire paper and then reflect at the very end. I try to bring out for them that the reflection is the glue that binds the exploration. You did this, what did you learn? How does your new knowledge lead you to your next calculation?

The abstract (§6) is not required by the IBO, but I find it extremely helpful at the end of the process to have the students write down exactly where they feel they have achieved the goals of the last three criteria. Anything they have missed becomes readily apparent in writing the abstract. It is also quite helpful for conferencing because in a paragraph or three you can quickly see where they need help without reading an entire draft (and you are only allowed to read one draft (HL Mathematics Guide, 2012, p. 64; SL Mathematics Guide, 2012, p. 43)). In addition, abstracts are required for the extended essay, and this gives them practice.

Polishing the exploration (§8) is mostly about helping students to ensure they have all the easy points and should come at the end of the process. I like to have all the students complete their initial exploration in year 1, and then if they have struggled, I give them the opportunity to do it again in year 2 on a different topic (it must be a different topic because only two complete drafts (§7) are allowed to be read by the teacher (HL Mathematics Guide, 2012, p. 64; SL Mathematics Guide, 2012, p. 43), but sometimes the students are really struggling because of the topic they chose so the fresh start is good from that perspective anyway).

I hope you and your students have a pleasurable journey through the exploration process because the most important lesson you can teach your students is to enjoy the mathematics.

-Daniel D. Slosberg

NOTES TO STUDENTS ON THE EXPLORATION

The mathematical exploration is your chance to use the math you have been learning so far in the course in a way which is meaningful to you personally. You see, all of the math you will be learning--calculus, statistics, probability, trigonometry--it is a collection of tools. They are powerful tools, and you may have mastered each and every one of them, but unless you use them in a meaningful way they sit in the back of your mental closet collecting dust.

This is your chance to explore math which is related to your own life, passions, and interests. It is also a time in the course when you can explore the ethical, social, or moral implications of mathematics, its applications and associated technology, or the international dimensions of mathematics (SL Math Guide, 2012, pp. 66-7; HL Math Guide, 2014, pp. 66-7). Through the process of writing your exploration, you will have the chance to first explore your interests, then to explore the math related to those interests, then to apply that math to your area of interest to solve your own problems showcasing your correct use of mathematical notation in your presentation, then to polish your work into a clearly communicated essay. It is likely that as you explore you will initially model your situation in a simplified manner. Upon reflection, you find more real world complications to add to your mathematical model. In the end, you will reflect further on how well the math has helped you understand the real world situation and what further directions you could explore if you had more time. The IBO will be assess the five areas mentioned above in the five criteria: communication, mathematical presentation, personal engagement, reflection, and use of mathematics. The purpose of this short book is to help you gain the highest mark in each of these criteria.

Once you have completed your exploration, you will have experience applying math to your own life, and my hope is that you will choose to do so on a daily basis. Mathematics is a powerful set of tools in your mental toolbox, and the course you are taking will help you to hone your mathematical skills; but unless you choose to apply your math to the problems you face in your life, it is a useless tool. Enjoy your mathematical exploration.

Mr. Slosberg

CHAPTER 1
BRAINSTORMING

Finding Your Topic

    Pick a Focus
    Brainstorm Real World Problems
    Brainstorm Related Math
    Choose Individually

To begin preparing for your internal assessment, you should begin with a stimulus or focus around which you can brainstorm how math is relevant to your life. It would be good to check with your teacher at this point whether they have a specific stimulus for your class or whether you are free to pick one on your own.

On a large sheet of paper, put your focus in the center. Perhaps you have chosen water. Then around your focus write as many real world activities as you can think of: water skiing, clean water, water conservation, rain, snow, ice, crystal structure, surface tension, etc. Around each of the real world activities, write the math that you believe might be used to understand them better. Geometry or fractals might go around snow flakes or crystal structure, for instance. This type of brain storming is best done with a few friends. Your teacher may set aside a class period to work on this in groups, or you could do it with your study group.

Once the mind map is complete, you are looking for a topic with three characteristics: it must be personally relevant to you (this is criterion C and is worth 4 points), it must contain math which is at the level of your course or slightly beyond it (this is criterion E and is worth 6 points), and it must be easy enough that you are confident that you can do it correctly. Criteria C and E are worth 10 points of the 20 points for your IA, so taken together choosing a good topic is worth half your IA grade.

One final point, each person in the class should have a different topic. It is fine to create the mind map together, and to bounce different ideas off each other, but make sure you and your friends are each working on different topics in the end.

Mr. Slosberg

CHAPTER 2
PERSONAL SELECTION
CRITERION C

Personal Engagement is 4 Marks

> Write in First Person
> Relate to Your Own Real Life
> Create Unique Examples & Claim Credit
> Cite Research in Bibliography & Footnotes

The exploration is a personal affair. Personal engagement is worth 4 marks while mathematics is worth 6 marks. Think about that for a moment. If you have very good personal engagement, but your mathematics is too simple, you have 4 marks for personal engagement and 2 marks for math for a total of 6 marks. If you have excellent mathematics, far beyond the level of HL Math, but you lack personal engagement, you have the same 6 marks.

So how do you get personal engagement? It starts by having a topic that truly interests you. It could be something from your past that you can connect to stories about yourself, or it could be something from your future that you are preparing for with good reason. In any event, you should write in the first person why you chose the topic and relate it to your own life with examples.

As you explore the mathematics, write about the books you read, or the websites you researched, and how you struggled and overcame your struggles. Make sure you footnote all of your sources and include them in your bibliography as well. Many students make the mistake of researching and writing first, and trying to remember at the end where they got their information. It is much better to footnote right from the beginning. If you don't want to put in all the details, just put the name of the article and the page number so that when you go back later to fill in the details, you will remember which source you used.

When you do the math, make sure you create your own examples. Make sure you mention that they are your own examples so that the grader is aware. Choose creative examples or find data on your own to use. Always tell the grader in your essay how you gathered your data so they know it is your own (i.e. that you were personally engaged) and not simply downloaded from the internet (which, if you haven't cited it, would be plagiarism).

If you have created a computer program as part of your exploration, it is important to show that it was a program you created (showing a lot of personal engagement) and not a program you downloaded ready made from the internet (which doesn't show as much personal engagement). One way to do this is with a GitHub (Yong 2015). When you create and modify your computer program, you commit changes as you make them, and this documents your learning journey and allows the grader to run the program you made as it was at various stages (useful if you made correct conclusions from incorrect data due to a mistake in your code). In your exploration, where you show snippets of your code, you should reference the GitHub so that it is easy for the grader to award you with the personal engagement you deserve for your efforts.

When you transition from one problem to the next, reflect on what you learned from the first problem and why you have chosen to progress to the second problem. These should read like your personal insights (because they are).

Finally, your conclusions should be your own. Say why you think your conclusions are correct and what you would do in the future to follow up.

CHAPTER 3
MATHEMATICAL SELECTION
CRITERION E

Mathematics is 6 Marks

>Do Math on Paper First
>Type Math into the Computer
>Proofread Each Step
>Avoid Copy & Paste Math Errors

One of the hardest parts of doing the exploration is choosing your mathematics. Your mathematics has to be of a high enough level for your course, but also has to be legitimately linked a personal interest or passion of yours. Finding something you are interested in and doing some math related to it is easy. Finding math of a high enough level and doing it is similarly easy. Finding math of a high enough level dealing with a topic that you are personally engaged with is a challenge and may take experimenting with a few different topics before you settle on one which is right for you.

You want to make certain the math you choose is of a high enough level to earn you a good mark. If you are in HL Math, simply make certain some of the math you are doing comes from the higher level portion of the syllabus and not from the core common to SL Math. If you are in SL Math, you want to make certain the math you choose is not part of the SL Math Studies course.

Once you have your topic and your mathematical connections, I recommend doing the actual math on paper first. Then, as you type the equations in and explain them, you can check your calculations and make certain they are right. It also becomes clearer which terms and variables you need to define if you are defining them as you type in the equations which use them without worrying about how to do your calculations at the same time.

It is tempting to copy and paste large portions of calculations and then to attempt to edit them. There are two reasons not to do this. First of all, your exploration should be about you learning and applying math and explaining it to someone else. Each type of calculation should be done once, but once you have shown someone how to do it, going through all the steps again gets repetitive and boring. All you need to say is that through the same process you got your further results. Don't repeat the calculations.

Secondly, if you have copied equations that you really do need, it is very easy to forget to change something, especially in a long and complicated calculation. If you don't change everything you need to, then your new equation is simply wrong which loses you marks on criterion E (especially if you are in higher level math) and, depending on what is wrong, possibly on criterion B as well.

When you have finished your paper, it is important to double check all the calculations. I recommend printing out the paper and checking that the calculations are right by redoing them in the margins. Remember, in criterion E in HL Math, levels 3-5 require that the mathematics explored is correct and level 6 requires the mathematics explored to be precise which the IBO defines as error-free to the appropriate level of accuracy (IBO HL Guide 2014, p. 70). SL Math is somewhat less stringent, but still requires correct mathematics for a level 6 (IBO SL Guide 2012, p. 49).

CHAPTER 4
CRITICAL REFLECTION
CRITERION D

Reflection is 3 Marks

Reflect Between Each Calculation
Summarize Reflections at the End

By this point in your mathematical exploration, you have chosen a topic and begun doing math on paper and hopefully started to write up your work. You have probably done research in several sources including your textbook, websites, other books, and articles. Make sure you have a title page and that you have sections titled: abstract, introduction, aim, rationale, body, reflections, conclusion, and bibliography. Creating these sections now--even if you don't have something to put in every one yet--reminds you as you are working on your essay to be thinking about what should go there.

Full marks on criterion D requires "substantial evidence of critical reflection" (IBO SL Guide 2012, p. 48; IBO HL Guide 2014, p. 69). In many ways, the body of your exploration is a series of math problems. Each time you finish a math problem, you should reflect on what you have learned by doing that problem and consider how that knowledge has led you to the next calculation. This is also a good place to get points for personal engagement by making your response personal and talking about how you decided to proceed (i.e. I decided to create another example using a larger block of ice -- say an iceberg). You get points for personal engagement by responding in a unique way, creating your own examples, and generally being creative.

Your final reflections sections should be critical of what you have done. You don't need to put yourself down or say your work was bad, but talk about how it could have been improved, for instance "I have assumed that Pluto's orbit around the barycentre of the Pluto Charon system is circular. If the actual orbit is elliptical, which is most probably true considering Kepler's Laws of Planetary Motion, then my calculations using a simplified circular model will have a few distinct errors" (Leung 2013, p. 12). They should also give an indication of further avenues to explore.

Mr. Slosberg

CHAPTER 5
TEACHER CONFERENCES

Teacher Time is Valuable

Prepare Specific Questions in Advance
Highlight Personal Engagement & Reflection

The IBO recommends only 10 hours of teaching time are spent on the exploration. If your teacher gives you an opportunity to do a second exploration, that means each exploration only gets 5 hours of class time. Some of that is spent teaching you how to write an exploration and guiding the class towards fulfilling the criterion. If there are 3 hours left for conferencing and there are 20 people in your class, you have less than 10 minutes to conference with your teacher. If you just show up and say "how can I make my exploration better," you are not going to make very effective use of the time you have with your teacher.

The most important questions you should have are math questions. If you get stuck somewhere in the math, frequently a few words from a teacher can put you on the right track where you otherwise might spin your wheels for several hours. Teachers also may know a good book that talks about your topic so you can get more information quickly and easily. While your teacher can point out errors, you have to correct them yourself. (IBO HL Guide 2014, p. 67). Flag the areas where you are stuck, indicate which results you want your teacher to check, and note where you need more sources of information.

The second piece you want to talk to your teacher about is personal engagement and reflection. I recommend taking two highlighters of different colors and highlighting your personal engagement in one color and your reflections in another color. This will help your teacher comment on them easily. You can also then quickly spot places where you don't have enough reflection or personal engagement--even before you talk to the teacher.

If you have any specific questions about your topic or the process, you want to make sure you write them down so you don't forget them.

Finally, as you approach the end of your exploration process you should write an abstract. This is an excellent tool for conferencing with your teacher since by reading your abstract your teacher can quickly figure out where you are having trouble in your paper in meeting the criteria. It also helps the grader find and award you all your marks whether that grader is your teacher or the moderator.

CHAPTER 6
THE ABSTRACT
CRITERION A

Communication is 4 Marks

Tell Them What You're Going to Tell Them
Tell Them
Tell Them What You Told Them
Abstract: Grader, I Deserve Full Marks Because…
Rationale & Aim Subsections in Introduction
Reflect Everywhere
Include a Reflection Section
Conclude after Reflecting

My aim in this section is to teach you how to create something which is not required for your internal assessment.

My rationale is that good communication in your mathematical exploration is worth 20% of your exploration grade and writing a good abstract will not only enhance your communication, it can make it immediately clear to any grader--be it your teacher or the moderator--that you have fulfilled the rubric and deserve full credit. More importantly, by writing a good abstract, flaws in your exploration will become readily apparent and you can fix them before turning in your all important rough draft.

To introduce you to the idea of writing an abstract, I would like to present to you the abstract of one of my students named Jessi. Jessi investigated wait times at McDonald's and produced the following abstract for her exploration. I have put key words in bold and underlined them for reference in this discussion, although they were not in bold or underlined in the original.

JESSI'S ABSTRACT

In this internal assessment, I looked into the waiting time, from queuing to placing order, at the fast food restaurant McDonald's. I stated in my rationale that McDonald's promise is to serve their customers within 90 seconds. Since I visit this restaurant often but always wait for more than 2 minutes in the queue, I aimed to explore how well McDonald's keeps their promise of customer waiting time in the queue and to find the probability that customers do not have to wait as long as I did for their meal. The use of mathematics from the HL core is the binomial distribution. From the statistics option, which was not studied in class, I applied 10% and 1% hypothesis testing and calculation of the chi-squared distribution. I began by stating an assumption of a binomial distribution's parameter and used hypothesis testing to determine whether it is true. Then, using the accepted parameter, I concluded that the null hypothesis, with probability of waiting time being less than 2 minutes as 0.6, is rejected using 10% significance level testing, but accepted in the 1% significance level. Reflecting on this outcome, when my alternate hypothesis is accepted at 10% hypothesis test but not accepted at 1%, I can be 90% certain that it is not true that the probability of waiting less than 2 minutes is exactly 0.6, but I cannot be 99% certain because there is not enough evidence to reject null hypothesis at this significant level. Taking into account the variables and uncertainties in real situations, I concluded that I should accept the results of 1% hypothesis test and the probability of waiting less than 2 minutes as 0.6 because there lacks evidence to reject my null hypothesis at 1% significance level.

As I sat in McDonald's and collected data, I considered different types of error that I may be making and real life scenarios that may affect the results. This experiment has many restrictions, such as small sample size, inadequate number of groups observed, and only collecting data at one outlet, which is unfair as the results are generalized to all outlets. To achieve a more accurate investigation, improvement can be made by increasing the number of customers to observe per group, observing more groups for chi-squared testing and recording observations in different McDonald's outlets.

(Lui 2013)

ANALYSIS OF JESSI'S ABSTRACT

Let's analyze Jessi's exploration through the eyes of the criteria. Analyzing criterion A, she starts out introducing her topic, clearly giving her aim and rationale, and she states her conclusion. You can immediately see in this first page that she has satisfied the requirements for "a well-organized exploration" as she "includes an introduction, has a rationale [...], describes the aim of the exploration and has a conclusion" (IBO HL Guide 2014, p. 68). In terms of coherence, you can see her main arguments in enough detail to decide whether you agree or disagree. (She decided to accept the null hypothesis because she could not rule it out with 99% certainty even if she could rule it out with 90% certainty. If you were reading the essay, you might decide that for yourself that you might have rejected the null hypothesis on the strength of the 90% test instead, but you can understand why she made a different choice.) In terms of completeness, you can see in her final paragraph many of the errors she considered and you can decide for yourself how complete you feel the list is. The abstract, which should always be less than a page, is necessarily concise whether the paper is or not. By the time the grader has finished reading the abstract, they have a sense that Jessi's paper will be coherent, well organized, concise and complete--all the things needed to earn 4/4 on criterion A. Now the rest of the paper still needs to live up to the standard created by the abstract, but by writing the abstract Jessi has put herself ahead of the game.

Jessi will still need an introduction including subsections on aim and rationale, but having included them in the abstract the grader can already check them off the list. The key principle here is "tell them what you're going to tell them, tell them, then tell them what you told them." You don't want a grader to have to search through your paper to find marks for you, you want to make your marks as clear to the grader as possible. The abstract is a summary of your entire paper. The introduction tells people what you are going to talk about in your paper. The body of your paper tells them what you want them to know. The reflection section and the conclusion section tell them what you've told them already. By following this structure, you will have a very clear essay.

In terms of criterion C, you can immediately get a sense that Jessi is personally engaged. She writes in the first person and describes the real life reasons that she was personally interested in this particular mathematical problem. She has made the problem her own by gathering her own data from a real life context.

In criterion D, she specifically reflects in coming up with her conclusion in the first paragraph, and then ends the abstract with a second paragraph dedicated to reflecting on possible errors. You have a sense that throughout Jessi's paper you will be able to find reflections as well as in the conclusion.

An exploration is, in many ways, a series of math problems that you do. As you structure it, the reflections are really the glue that holds your paper together. After each problem in the paper, you should reflect on what you learned by doing that problem, and reflect on what you are about to do in the next problem and what you hope to learn from that. An abstract is quite short and cannot hope to contain all of these reflections, but the most important ones should be included in the abstract.

Finally, before your conclusion section, you should devote an entire section of your explorations to your reflections. Summarize the reflections you have already made and reflect further on how you are deciding on your final conclusion and what you have not completed--further investigations that could be made or errors which could not be resolved. You get a good sense of what this might look like in Jessi's final paragraph.

In criterion E, she points out up front which mathematical ideas come from the HL core material as well as which come from the HL options. This clearly demonstrates that the work is commensurate with the level of the course.

You will notice that the abstract allows you to address every criteria except for criterion B. It is not appropriate to define key terms in an abstract, or to use multiple forms of mathematical representation. Certainly you should use appropriate mathematical language in your abstract, but it is more important that the language remains appropriate throughout your paper.

Not only is an abstract good for addressing the criteria, it is very useful in your conversations with your teacher while you are preparing your exploration. Your teacher can quickly identify holes in your arguments because they do not show up in your abstract. Your teacher can also point you to criteria you may have done poorly in because they look poor in the abstract.

If you reflect that a class with 20 students during a 60 minute period only allows 3 minutes of conference time per student, you will understand that if you hand your teacher your entire paper and say where should I improve, they won't have time to read the whole thing. If you have an abstract, however, they can read your abstract, quickly find the potential problem areas in the paper, read those sections, and they can help you quite a bit in those three minutes.

In conclusion, an abstract is a wonderful way to demonstrate to any grader--whether they are your teacher or a moderator--that you should have full marks on your exploration from the first page that they read. It enhances your communication by being a concise and complete summary of your well organized and coherent essay. It showcases your personal engagement, use of mathematics, and your reflections. In addition, it can be a valuable tool in conferencing with your teacher to ensure that your teacher focuses on giving you help in the areas where you need it most.

Because it is the first thing the grader sees, make sure you spend extra time proofreading the abstract to make sure you make the best first impression you can. Spending extra time proofreading your abstract is especially important because it is the last thing you write and therefore you will not have been editing it as you worked on the rest of your essay.

Mr. Slosberg

CHAPTER 7
ROUGH DRAFT

Remember: You Only Get One

Finish All the Math
Reflect After Each Calculation
Personal Engagement in Introduction, Rationale, Sample
    Problems, Reflection, & Conclusion
Double Check Notation
Communicate Clearly
Polish Your Abstract

You can ask your teacher about your essay as many times as you like while preparing your rough draft. You can show your teacher various sections and ask about the math, or how to bring out the personal engagement more, or if you've missed anything in your reflection. The abstract is a wonderful tool for you to use in conferencing with your teacher because they can quickly pinpoint the places which need work in your essay and they can flip to that part of your paper and discuss it with you. These are all useful things, but eventually you will want to show your teacher your entire essay. You are only allowed to do this once.

Check with your teacher early on about what internal deadlines they have set for you, including the date your rough draft is due and the date your final draft is due. Because you only get to submit one rough draft for them to read all the way through, you want to make sure it is very good. Keep in mind, your teacher is only allowed to write short comments and point out errors -- they can't annotate it thoroughly or correct your mistakes -- so you want to have it as close to your final draft and as polished as you can make it.

You should have all of the math completed by the time you submit your rough draft. Your teacher can only find errors if the work is complete. If you are wondering what further problems you could work on, make sure you ask your teacher long before the rough draft deadline so you have time to complete the work before you turn in your rough draft.

Every time you finish a set of calculations, take a paragraph to step back and analyze what you have learned from that calculation and to segue into the next calculation with your reasons for heading in that direction. These reflections are a chance for you to think critically about how robust your calculations were. In your initial problems, you probably made simplifying assumptions which helped you do the math. This is your chance to talk about them and talk about which ones to get rid of in the next calculation. You can also talk about how accurate you think your calculations are, and how you might make them more accurate.

At the end of your paper, before you begin your conclusion, you should have a section reflecting on your entire process. Think about all the calculations you have done and everything you have learned from them and write it down. Think about what the next steps might be if you were to continue.

Go through your paper very carefully checking your notation. If you are writing something which not exactly correct because you are rounding a decimal to three significant figures, for instance, make sure that you use an approximately equal to sign such as $1/3 \approx 0.333$ to avoid being penalized in notation. It is very easy when copying equations and manipulating them in a word processing document to forget to change something small, so it is helpful to redo your calculations by hand in the margin of a draft of your paper to check each step and make sure you haven't made a typo.

Good communication is worth 4 points, so you want to make sure you paper flows smoothly. When your rough draft is almost ready, it would be a good idea to show it to a friend and to your English teacher or some other non-math person to get feedback on how it reads. If they get stuck somewhere, it is probably because you haven't given enough explanation for them to understand where you are going. One typical error is to have a list of equations without explaining what you are doing in each step. Compare

$$\int \frac{2x}{x^2} dx$$

$$u = x^2; \; du = 2xdx$$

$$\int \frac{du}{u} = \ln(u) + C$$

$$= 2\ln(x) + C$$

with the following:

I would like to demonstrate the use of u-substitution by showing that

$$\int \frac{2x}{x^2} dx$$

equals $2 \ln(x) + C$. First of all it should be clear that

$$\int \frac{2x}{x^2} dx = 2 \int \frac{dx}{x}$$

and given that

$$\int \frac{dx}{x} = \ln(x) + C$$

from our data book, we know

$$2 \int \frac{dx}{x} = 2\ln(x) + C$$

so you can see that what we're trying to prove is, in fact, true.
Now if we let $u = x^2$ then its derivative is $du = 2x\,dx$ which is very conveniently the numerator of our integral. Substituting in u and du, we have

$$\int \frac{du}{u} = \ln(u) + C$$

We can now plug u back in to get

$$\ln(x^2) + C$$

and by using our laws of exponents we have

$$2 \ln(x) + C$$

as required.

All the math in these two examples was the same, but you can really see how much more smoothly the secondly explanation flowed compared to the first one. It doesn't take a lot, just adding a few words of explanation between each line of equations increases the readability of your work quite a bit.

Once you've checked that all your equations are correct and are well explained in the body of your exploration, go back through and look at each figure and table. Have you written a title centered above each one? Have you put a caption with a few sentences of explanation below each one? Are they numbered sequentially? If you cover up the rest of the essay and just look at the title, caption, and figure, can you understand what it is trying to say? Have you referred to each figure somewhere in your text, preferably on the same page as the figure?

Once you feel your paper is in good order, turn to your abstract. Your first sentence should give the aim of your exploration. Make sure you use the word aim in your sentence. The second sentence should give your rationale showing clearly how you are personally engaged with the topic. Be sure to use the word rationale in your sentence. Next you should indicate the math you have used and why you feel it is commensurate with the level of the course. In two or three sentences you should be able to explain everything you have done. At the end, you should showcase your main reflections. Once again, use a form of the word reflect in this sentence. Finally, in a single sentence you should summarize your conclusion.

When you feel everything is ready, turn in your rough draft and take a break. Once the teacher returns it, you will have more work to do.

CHAPTER 8
POLISHING
CRITERION B

Notation is 3 Marks

Define Terms
Use Equations, Diagrams, Tables, & Graphs
Check All Equations
Check Each Step
Check Approximately Equals Signs

The final stage is about checking that what you have done you have done correctly and completely. During this stage, you will proofread your essay many different times in many different ways.

The first thing to do, of course, is to fix anything which your teacher pointed out was wrong in the rough draft. Make sure you understand each and every one of your teacher's comments. Because teachers aren't allowed to make very many comments on rough drafts, each one is important. Go back and ask the teacher if you are unsure of any of the comments.

The next thing to do is to go back through the paper making a list of each specialized term (for instance boat speed) and make sure you've defined it. (did you mean the speed of the boat as compared to the water or compared to the land? For the same boat, these two values will be different on a river.) Do the same for each and every variable. Every letter in your paper should be defined either mathematically ($u = x2$) or in words (s is the boat speed, i.e. the speed the boat is moving compared to a fixed point on the land).

After that, consider all your arguments. Have you shown each point as well as you possibly could? Would a graph have made things clearer? A table? A diagram? Diagrams can frequently be very helpful in defining terms. Tables can be helpful in summarizing results without repeating the same calculations over and over again. Graphs display conclusions clearly. Make sure each figure--whether it is a graph, table, or diagram--is clearly labeled with a title above it and a caption below it. It can be helpful to cover up the text of your essay and show the title, figure, and caption to a classmate. Ask them if they understand why that figure is in your paper just based on the title of your paper and the title and caption of the figure. If they don't, that probably means your caption isn't clear enough. While you're looking at figures, check that you have units everywhere they are needed (chart headings, graph axes, diagrams).

If you have taken a graph or other figure from a computer program, make sure all the notation is correct. Perhaps the graphing program you are using shows r^2 = 9.937643728048E-01 on the graph. This is not correct notation. In your caption you should write something along the lines of "The correlation coefficient for this graph is quite good ($r^2 \approx 0.994$)." This shows the grader that you understand the notation is incorrect and know how to write the correct notation.

Check all your equations. Make sure you have written what you intended. An easy way to catch copy and paste errors in equations is to write two equations on the top and bottom of a piece of paper and work out how to get from one to the other in the middle. If you can't get from one to the other this time, you probably typed one of the two equations incorrectly.

Finally, check any place you have a number. In the old notation criterion, the fastest way to lose a mark was to use = incorrectly instead of $\approx$ in any two places in your paper. The rule is to always use $\approx$ when you are rounding a number, for instance $\pi \approx$ 3.14. If you are in HL Mathematics, remember that "precise mathematics is error-free and uses an appropriate level of accuracy at all times" (IBO HL Guide 2014, p. 70). This means that getting a level 6 in criterion E can be put in jeopardy by not paying attention to the appropriate number of significant digits to include. It is rarely appropriate to simply write down the answer as displayed on the calculator.

BIBLIOGRAPHY

International Baccalaureate Organization (IBO). HL Mathematics Guide. August 2014. Pp. 64-70.

International Baccalaureate Organization (IBO). SL Mathematics Guide. March 2012. Pp. 43-49.

Leung, Rahim. "Determining and Applying the Scalar Gravitational Potential of the Vector Newtonian Gravitational Field." May 20, 2013.

Lui, Jessi. "To Explore the Waiting Time, from Queuing to Placing Order, in a Fast Food Restaurant." May 20, 2013.

Yong, Cho Yin. Personal communication. June 30, 2015.

Mr. Slosberg

ABOUT THE AUTHOR

Daniel Slosberg has always loved math, from the time when he was four year old and gave his nickel allowance to a neighborhood child to teach him long division, to the time when he was 11 and used a spreadsheet to show his parents why the clothing allowance was not lost but had merely been transferred to the computer allowance column in his ClarisWorks spreadsheet: a temporary situation involving negative numbers which might be remedied at some future date. His father started teaching him mystery math (algebra) in primary school from the encyclopedia because math at school was too easy. He joined the Study of Mathematically Precocious Youth run by Johns Hopkins University for scoring over 600 on the SAT-M before the age of 13, and he scored a perfect 800 in 8th grade. At a restaurant one evening, he asked his father how computers think so his father taught him base 2. He had figured out bases up through 11 by the end of the meal. By grade 6 he was 1st in the State of Michigan in math and tied for first in the United States and by 9th grade he was taking AP Calculus. In high school, he represented the United States in the American Regions Math League's trip to Russia. Tenth grade saw him taking university math, and university saw him taking graduate level mathematics and physics.

Daniel started teaching in 1995. Astronomy at first, and then physics, math, computer programming, architecture and various other courses to the present day. He has taught both HL & SL

Mathematics since 2004 starting in Centennial High School in Corona, CA and continued in Hong Kong, China since 2007, working his way up to Head of the Mathematics Department at Victoria Shanghai Academy (in Hong Kong, not Shanghai). It is his hope that this short book will allow both his own students and students all over the globe to achieve greater success in writing their mathematical explorations so that they will be under less stress when they write their external examinations.

The future, however, is why Daniel writes. He feels every young person requires a strong foundation in mathematics, not for its own sake, but to be used throughout life. He hopes the students reading this book will enjoy the mathematics in their explorations and will apply math usefully in their daily lives long after they have graduated from the diploma program. If you truly understand compound interest, you will be rich regardless of how little money you make in your job. If you buy your first stock at age 16, you are late, but not too late. Personally, Daniel will use his math to figure out how to live on Mars--you can read his method for extracting water from Martian soil first published in the Proceedings of the Mars Society and now available on Kindle as The Martian Farmer.

EARNING FULL MARKS BOOKS

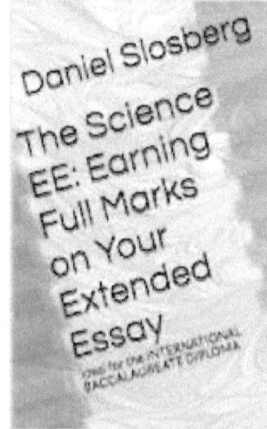  

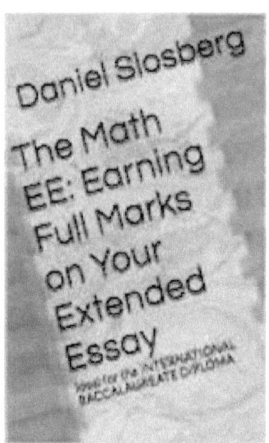 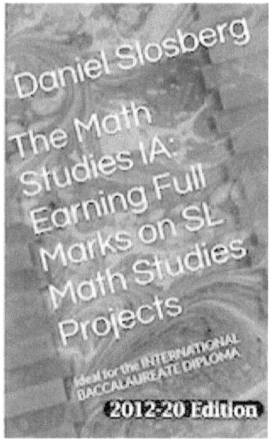 

Rainbowdash Publishers LLC is currently seeking teachers and 1st year university students who earned 7's in DP to author books for additional courses, especially in groups 1, 2, 3 and 6. eMail RainbowdashPublishersLLC@umich.edu if you are interested.

MR. SLOSBERG'S OTHER BOOKS

OTHER RAINBOWDASH AUTHORS

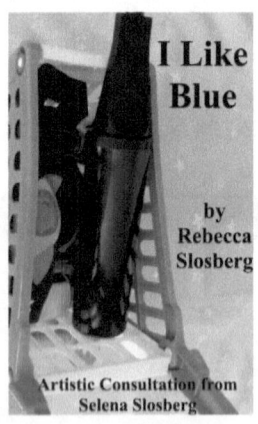

www.ingramcontent.com/pod-product-compliance
Lightning Source LLC
Chambersburg PA
CBHW020956180526
45163CB00006B/2388